The Cat Chat Forum

The Cat Chat Forum

1st Edition

Copyright 2013 By Laura L. Smith

No part of this book maybe reproduced, copied or electronically transmitted without expressed written permission from the author. Disclaimer: The views presented in this book are from a cats mind and we cats assume no responsibility if the reader doesn't find how we think funny, interesting or educated. No cats were physically hurt or psychologically damaged during the photography sessions for this book. And we were all provided with proper feeding and watering afterwards too.

All photographs and paintings by Laura L. Smith
Book cover photo: Mountain lion
Inside cover photo: Dreaming of you

Contact the author at: laural.s2012@gmail.com

List of Photos in the book

Walking the tub line	4		Cat and dog stare down	33
The disgruntled black cat	5		Cat flipping the paw	34
Mistrustful dog and cat circling	6		Two cat standoff	35
Dog and cat trash talking	7		Taking the cat food dish	36
Angry cat waiting by the door	8		Yawning cat	37
Cat dropping in uninvited	9		Narcissistic cat looking in mirror	38
Indifferent cat eating from bowl	10		ADHD Kid riding dog	39
Sissy cat in lace	11		Cat toss	40
Angel Cat painting	12		ADHD Kid hugging cat 1	41
The Cat Lady painting	13		ADHD Kid hugging cat 2	42
Kitten stuffed in black shoe	14		Cat coming in door	43
Entwined adult cats	15		Cat walking on fish tank	44
Napping kittens	16		Cat sniffing cheap wine	45
Preening cat licking his paw	17		Cat watching TV show	46
Waiting for Dinner painting	18		Cat with foot in front of him	47
Kitten looking at dogs tale	19		Cat dreaming	48
Cat looking at foot	20		Man with his cat by tree	49
Mad cat on towel rack	21		Cat with roses	50
Cat over light	22		Cat Christmas card	51
Cat napping on chair legs	23		Disgruntled morning cat	52
Cats playing in shopping bag	24		Cat and mouse card	53
Cats sharing the milk	25		Psychedelic cat face	54
Kittens crying for mama	26		Psychedelic cat art	55
Cats ready to fight	27		Disturbed cat by food bowl	55
Kitten and dog introduction	28		Sleeping kittens in box	56
Cat and dog standoff	29		Man with his cat 2	56
Painted grumpy cat	30		When robins attack	57
Cat being petted	31		ADHD Kid hugging cat 3	58
Cat on woodpile	32			

Do cats chat? Of course they do. They were the animal that the Egyptians worshipped and painted on the walls of the pyramids, hence some of the cat paintings in this forum. So it only makes sense that these opinionated cat sayings, photos and captions are presented here. So enjoy what they got to say, after all it's the wisdom of the cat that makes the day.

You know it's a sad day when I have to go outside and meow to the neighbors what I am thinking because my owners aren't listening. We cats have exquisite taste in fine foods, but that does not mean I want to eat pizza. So why is it that when I request cat nip, a pizza delivery guy shows up. Also, it seems that for some reason my owners think I am open to eating the meat off the pizza, and place it in my bowl. Well, now I have tainted cat food. Please, I have a sensitive digestive system and prefer a can of good quality cat food. Or if you don't have that available, tuna juice over my dry food will do just as well. Hey pass on the information to the rest of your human friends. Thank-you.

Here we are circling each other because we are trying to size up the situation. You know that it is true, we cats get our burst of speed, and can move quit fast. But I can't say the same for my human owners. They sit around a lot drinking beer, eating snacks, and watching the TV. No wonder there is an obesity problem in this house. They should get up and walk around, lose some of their fat. They could also maybe clean out the litter box a little more often and wash out the cat food and water dishes. And for the love of the cat, please put out those horrid cigarettes. We don't need to inhale second hand smoke and die from a lung disease.

Friendship between natural enemies? It is obvious that I am relaxed enough to lie down and lift a paw in a gesture of peace. Too bad that some of those people in foreign countries who are fighting over religion and political beliefs can't follow the paw lifting peace symbol. I bet that the homeless cats over there who are running around wondering where their next meal is coming from would love to extend their paws of peace. It's such a simple thing to do and doesn't take much energy, plus it tames the egotistical hot tempers that rule people's emotional states. Maybe <u>Raise the Paw; Love the World</u> should be the motto of the day.

Patience my paw. Kittens have a lot energy because we are growing. I am quite tired of sitting here and waiting for a human to open the door. Don't they realize that a cheaply made screen door is no match for my claws? Even though they are small, it only takes a swat of my paw to start a hole. Then the other adult cats in the house will finish ripping it open so we can just come and go as we please.

So let's forget about this patience stuff and just open the door. I haven't got all day to wait. The weather is nice; there are birds to kill and trees to climb. Let's get this cat show on the road.

Cat's are really not into company just dropping in. When a cat is the only pet in the house and the humans bring home a new cat, we don't take kindly to it. We tend to get upset and start marking our territory with our spray. We will mark the legs of tables, chairs, even bed linens—you know the drill because you smell it. Then you have to go buy cleaner to remove it. So please don't buy this garbage about friendship between us cats. It isn't happening. But then humans never listen to us anyway. They tend to do their own selfish thing and they certainly don't speak meow. I think there should be a meow law passed where all cat owners must learn the meow language.

I offer my poetic outlook on the above photo.

I am eating my food. Please do your own thing. If you wish to let your egos get the best of you, then leave me alone. Crunch, crunch, crunch is what I love to hear, as my teeth bite into each morsel of food. Such a heavenly sound. Oh the joys of eating while ignoring the growling and hissing. Such a beautiful thing, such a wonderful mindset of life. To eat my tasty food and not care about the silly cat fighting and strife.

Do I look like I fell for the same old trick again? Do you think I like to wrap up in some old lace cut off a ten year old dress that was worn to a prom by some human? Get real. The only reason I am doing this is because my owner's kid was holding up a can of cat food above me. Then my owner starts in with the goofy mushy talk. *"Okay kitty look up here. Oh such a cute kitty. Love you"*. Oh please, mushy talk will not win my cat heart of love. Food, catnip, rubbing my ears and petting me when I am in a good mood does. So just give that can of cat cuisine kid so I can eat. Gracious!

Title: Angel Cat. A multicolored acrylic painting on rag board. You know when we cats look at this we wonder; what was the artist drinking when she painted this? The idea is good though as we cats can be angelic, but what is with the gold wings and body colors? Cats have more cones than rods in their eyes for being able to see at night, so these colors are really vibrant, almost too much for my sensitive eyes. I hope I don't have nightmares about this one.

Title: The Cat Lady. Acrylic painting on rag board.

Wow, stop the presses. This is just too freaky. What's with the one eyed cats? Are they an alien experiment gone awry? And you call that a woman? She looks more like a waving blimp flying in the wind. And since when do cats hang upside down? Unless they are stoned. I know if I was being held upside down I would be ripping up the hands holding me. Let's move on before my mind accepts this cat art as real.

Okay putting a kitten in a shoe to make a point is a brilliant concept. The only reason it works is that the kitten is one of those Manx cats that has no tail. However, the captioning really does not apply to us cats. We cats are quite judgmental and we certainly don't have to walk a mile to make a judgment. It is a matter of our survival to judge our situations. We judge whether or not to eat the food by how it smells. We judge if our bed is comfortable enough to sleep on so we get our proper rest. We judge whether or not we will allow the humans to pet us so we don't get moody and scratch your hands. We judge the running distance to the closest tree when a dog is chasing us. Yep, we do judge every situation every day to stay alive.

Need I say more? I know that my buddy's owner thinks this is a cute photo, but it is really more emotionally deeper. My cat soul mate and I are both spiritual cats of planet earth who love each other with total dedication to the cause of peace and harmony. We find it great to express our love for each other in ways we find comfortable. We believe we are entitled to live what lifestyle we were born to live without human or other cat intervention. Cat peace be with you all.

Life is about living as a loving and purpose driven cat soul guided by God and our angels love. It is about knowing that love is the key to survival in the cat soul's spiritual universal plan. When we learn to love and let go of the fear, anger and negative ego—the blessings of the universe will manifest for us cats.

Life is to understand who we are and that love encompasses those cats who are around us.

We are love and love is who we cats are. Let love guide all cats to know the path on which they walk. We are never alone as we travel; the love of our cat angel does reside within our cat souls.

We cats really do not have privacy when we are living in a human household. Everything we do is out in the open. Think about it, where is the litter box? Is it in a cozy room where we can shut the door to keep other cats and human alike out? No. What about where we sleep? We can't shut the door and be frisky with another cat. We have to sneak outside to do that. And as far as Uncle Sam, I have heard his name mentioned several times when my owner's are talking about how he takes their money. But I have never see him because my owner's don't like and trust him. He sounds more like one of those cat relatives of mine that only come around when they want my cat food.

This fine painting is titled: Waiting for Dinner. Note the aquarium setting of swimming orca whales through the glass. Well, this is just pure fantasy. We cats would never eat an orca. They are just too big. It's more like the other way around; we would become an orca cat appetizer. When we do eat fish it is usually tuna from the can. Dry and canned cat foods that claim to have the flavor of fish, oh please, what a bunch of bunk. There is more fish flavor in those omega 3 oil pills that the humans take for their heart health. Oh by the way just ignore the cat floating in the corner. I think the artist was high on something and hallucinating when she painted it.

Such a sad state of affairs using this kitten to get a photograph. Humans don't place their babies in precarious situations, why do they think it is okay to do it with a kitten? That is a tail that belongs to the owner's 75 pound black Labrador. You know how kittens are, they are curious and playful. An adult cat will let a kitten play with their tale, but a dog? Luckily the kitten did not jump on the tail. Too, the dog was lying under the chair so it would have been hard for him to jerk his head around and snap at the kitten.

Humans must think we cats are stupid. I think what is stupid is the throw rug that will slide when the human steps out wet from the tub onto it. Then the human ends up falling and hitting their head and their out for the count. Then when they come to they start yelling, "I've fallen, and I can't get up". Well, we cats can't do anything. I don't know how to dial a phone. All I can do is go in and comfort the human. What worries me is if they aren't found my food bowl will become empty and I starve to death. Moral of this story: have a backup plan in case you fall. The life you save might be my own.

Yeah, I did not want a bath and I still got one. Then the humans stuck me up on the towel rack when I am still wet. Why do humans think we cats need a bath? That is what our course tongues are for. So what if they think I shed too much fur. Just give me that brown fur ball paste medicine on my paw and I will lick it off and barf up the fur. It is not that hard to clean up fur balls off the carpet. Hey God knew what he was doing when he made us cats; we have been surviving just fine for centuries without a bath. If we were made to go into the water then I would be out swimming in the owner's backyard pool.

This really isn't a tanning light, but I trust you humans get the message. I live in the desert and I like to lie in the sun in the cool winter time, but not the hot summer sun. I stay inside where it is air conditioned or find shade. Also I have fur that protects me from the ultraviolent rays. You humans do not. And no, heavy body hair on males don't count. Tanning lights are not safer; they damage the skin just like the desert sun. What gets me is that I will see my human owner light up a cigarette and open a can of beer and then go sit out in the summer sun by the pool. Can we say wrinkled and dry skin?

I love my simple stress free life taking those cap naps in the midday sun coming through the windows. I haven't much to complain about this photo. I slept right on through the noise of the shutter clicking. It was just me and the one owner. The kids were in school, and the other cats were taking their nap someplace else in the house. Peace and quiet—I love it. Except, of course, when the owners start fighting, the TV is blasting 20 hours a day, the kids are making noise, the cat dish becomes empty and the litter box is full of poop and not cleaned. Then I get to be a real grumpy cat. Growl, growl! Give me back my peace!

You know shopping is an obsession for some felines. My sweetie here likes to go with the owners to those pet stores that allow humans to bring in their pets. I don't go because the large crowds cause anxiety for me. When my sweetie views the large selection of cat stuff on the shelves she starts meowing and pawing furiously at the door. The owner then speaks that goofy baby cat love talk to my sweetie and loads up the shopping cart with more crap we don't need. Pardon the language. Of course, my sweetie is spoiled and I have to deal with it. Oh well, since I am a real tom cat and love her I guess I can forgive my lady for wanting so much.

Oh yeah we cats love sipping from a bowl of milk when it is at room temperature. But the truth is cow's milk is not good for us cats. It can cause some serious gas in our digestive tract. I know my gut will rumble as the gas is rolling through my intestines towards the exit. And when I let go, it can clear the room. My owner is running for the air fresher while waving her hand in front of her face. And of course where there is gas there will be smelly poop dropping out shortly into the litter box.

So the next time you are thinking about leaving that leftover bowl of milk on the table, remember when we blast out the gas, it's time for the mask.

This is not public speaking; it is kittens crying for their mama. The owner in this photo has these kittens up on a high table covered with a white slippery cloth. And what's with the fake flowers? But since we are talking about public speaking, I want to say that the post traumatic syndrome these kittens will suffer as a result of this photo shoot will cause these kittens to speak out against pulling them away from their mama when they need to stay close to her. But human's think it is okay to separate kittens from their mama when they want. Kittens are not chattel to be used for silly things such as photos. So let the kittens be; after all they have feelings too.

I like this photo. I do love a good cat stand off when warranted. But what is with the shoes? I bet they are smelly. Cats do have egos that can get out of control and get the best of them. Humans are really no different. I have heard and seen some good arguments between my owners. Oh they will stir up a storm with their cussing, yelling and hitting each other. But that is boring compared to when two cats are in a rolling brawl of paws and claws tearing into each other's fur and face. Then the loser runs off with blood dripping down their fur. But the winner doesn't walk off unscathed either. He might end up with infected boils that when they pop, green stuff squirts out. Oh well, you wanna fight, you gotta pay the price.

Tolerance of others should be practiced by all living creatures. We should not be so intolerant that it colors our view of the world. Again, that is the 75 pound dog and the kitten from an earlier photo. In this set up the situation is being supervised. You will note the kid has a good grip on the kitten, ready to yank it away in case the dog's tolerance wears thin. Labradors are known to be good family dogs, but again you don't want to push certain issues. And the above is one of them. Luckily the dog did finally accept the kitten. Warning: do cat and dog introductions under human guidance: you don't want to bury a dead kitten that died from a broken neck inflicted by a 75 pound black Labrador dogs powerful mouth.

I don't like bullies and I stand up to them. In some parts of the country I understand there are those neighborhood associations that fine humans with unruly pets. I have even heard through the cat grapevine about the humans being put in jail for letting their dogs bark endlessly. Then again dogs aren't as smart as us cats. In reality we cats make the dogs bark that get their human owners in trouble. One way is we will walk along a fence line slowly causing the dog to go crazy and bark. Or we walk across the yard when the dog is tied up or in the house, then stop to poop and not cover it. The owners haven't a got a clue it is us. Then Fido gets the blame. Ha, ha!

When cats go with the Gothic look

Even though this is just a photo painting where this does human get off thinking that it is okay to paint up a cat? Oh get real! Cats do not go Gothic nor do they wear makeup. The fur in the ears is even blue. Cats lived in the Gothic times, but did not partake of the drabby dress or darkness. No wonder this cat does not look happy. He looks like he is ready to bolt off and out the door. I don't blame him. I tell you if my owner tried to make me look like some carnival side show clown I would be spraying every piece of furniture in that house. These humans need to get a life and a reality check before they get swatted by a cat claw.

When cats get massaged

Alright, I love being petted. Petting is a good thing. It makes me feel appreciated. I love my owner's large hand gliding down my back, then him lifting his hand back up to my head and then back down.

I also love my ears being rubbed and my neck being stroked, gently of course.

One thing I do want to warn you about—do not rub my belly, I hate that and will tear my claws into your hands.

Other than that—happy petting!

There is that Labrador again. And what is with the cat on a wood pile? The lengths some humans will go to get a photo. And why do they insist upon having these pet stand offs? The dog has all the advantages. He is big, the cat is small and obviously terrified and unable to move. And this is called a creative photo?

However, once the cat gets back inside the house he will have his revenge by swatting the pooch's nose.

(By the way did I mention I hate poodles?)

Where is the human society when the cats need them?

Oh well, can we say stare down. Apparently, the cat is even more terrified. I assumed that the human would have shooed off the dog after the photo was taken. But no she sets down the cat and told the dog to lick the cat which got the cat mad and soaked at the same time. Finally the cat curled up around the dogs face and let him have it with his sharp claws. The dog jumped back startled and the cat ran across the yard and up the fence with the dog in hot pursuit.

What got me is the human thought it so creative that she made a postcard of the photo. That is so wrong human sicko!

A cat flipping the paw

This is not a happy looking cat. He looks depressed. Probably needs to speak to a cat counselor. Cat depression is a serious issue that needs to be addressed. Maybe this cat is not happy with the food or the sleeping quarters. Or maybe he feels rejected by the humans. And what about the litter box? He has to share it with three other cats. That in itself is not a healthy thing. And if this cat is not depressed maybe he has a physical ailment that needs a vet's attention. Maybe he is suffering from an digestive problem or has kidney or liver issues from too much ash in the food. All sorts of things. The best advice: Assess, assess, and assess your cat at all times. His mental and physical health depends upon it.

Humans can learn from these cats actions. When we cats find an argument is a losing battle we walk away. But what I don't understand is my human owners will keep arguing like six hungry Dobermans fighting over a can of dog food. Humans also yell out cruel words and hit each other. What a waste of energy. It is so much better to go take a nap and sleep off the anger. I do that and when I wake up refreshed from my nap, I hit the litter box and food dish. Then I wander outside for a touch of fresh air and sunshine. Think if the humans did that, there would be less wars and divorces. So the next time you humans are embroiled in some goofy conflict where no one is winning—stop and go take a nap. Nighty night!

Yep, self explanatory.

You can tell by the look on this cats face he did not appreciate his food dish being touched. Luckily this kid let it go.

I wonder what the mother was thinking when she planned this photo.

Obviously, the mother has never had her hand ripped up by a cat's claw when taking away their food. Using her son to do her dirty work is stupid. Good thing the kid survived without being inflicted with any cat claw scars.

Motto: if you are not willing to suffer the cat scathes then don't touch the food dish you foolish human.

Yes, this is an excellent way to make a human mad. Humans do not want to be taken lightly. They want to be heard. Even the nagging ones. My owner tends to be a nagger to us cats. Every time I turn around she is yapping away about how we are making a mess while using the litter box, eating or drinking. Well, come on lady, you got six cats and you're at work all the time. You can't work and expect to have a spotless house. If you want that then stay home or hire a maid. And seriously, nagging at your kids to clean up our mess is like talking to the wall. Hey, I can't get the kittens to do things right. Do you think it is any different with human kids? My advice? Clean it up yourself and stop nagging before you have a stroke!

It's a no brainer. Yes, we cats are genetically programmed to be narcissists. So please do not expect us to change. Because if you do, you're in for a long and disappointing wait because it's not happening.

I am not particularly fond of dogs but where is the parental supervision when a dog needs protection from the human kid with ADHD? However, the kid did not get bitten, just thrown off. Of course, the dog got away without any back damage to chase the family cats another day.

Cat toss is not a game. As you can see the cat is not having a fun time. Note the claws extended out of the paw. Amazingly the cat did not attack the girl when he came back down. A matter of fact he was so stunned by his toss up that he allowed her to pet him to calm him down. Moral: If you toss the cat he might lose his cat cookies.

Of course, share the love but don't strangle the cat. A little easy on the neck ADHD boy. And what's with the lower lip? It's the cat that looks like her air supply might be cut off. Moral of this ditty: Handle the cat with kid gloves, not a harsh hand.

Here is the same kid using his strangle hold on a different cat. The cat still does not look any better off. And since when is squeezing considered a loving hug? What I find amazing is the cat appears relaxed and almost bored. No claw extension out here. Maybe there is a trust factor between these two I don't' know about.

All I see is a good looking cat coming in the door. But then again since cats have souls, I guess they would have their own cat guardian angels. After all when cats die they go to cat heaven waiting for their owners to join them.

Critical thinking? The photographer wasn't using it when she left out the letter t in the captioning. I think those two words are a misnomer and over used. Critical means a dire situation. Sorry but my cat thinking is not a dire situation. It is innovative and inspirational. And besides the cat isn't going to get the fish without getting wet.

Cats have a very good sense of smell. They can tell when something is not good quality. Humans are notorious for being cheap and will buy a good brand dry cat food container and then put cheap bargain brand food in it. Hey you might get away switching out good liquor with cheap liquor, but not with our food. We won't eat it, so fageddabout it! On the other paw I have noticed when there are a bunch of human lushes partying; all they care about is the buzz. So any cheap brand of booze with do.

I guess it could be said that if the humans can't handle the cheapie liquor then don't come to the party.

Cats like to watch those nature shows that feature birds and wild cats. This is a good way to keep the hyperactive cat with cat attention wandering disorder also known as CAWD focused on something for longer than 10 seconds. However, when a commercial came on, this kitten started sliding down the slide over and over. Then she started playing with her tail and then ran into the cat food dish pushing it across the floor. Finally the human owners had to put her on medication. Cat Lord almighty, it finally brought us peace.

The human foot. It is meant for walking but humans have for centuries used their foot as a weapon to kick dogs and even their cats when they became mad. A matter of fact, cats have been kicked so much that it has become a genetic program in mind of the modern domestic cat. This ignorant human doesn't realize that I could have easily swatted her foot and making it bleed. But luckily since she is nice and treats me well, I just gave her a warning look.

*Now I lay me down to sleep...
Dream about those mice I eat
Think about cat nip I just can't beat
I hope someone doesn't want this seat...*

The truth of this cat poem rings true in the home. When a cat is sleeping they do dream about simple things. Birds, mice, food, and catnip. But there is also a darker side of fear that cats have they don't talk about. That is the inner fear is their human owner(s) will come along and disturb their sleep by picking up or moving them. This can cause the cat to be jolted awake leading to SCAT or sudden cat acute tachycardia. This condition can be prevented by leaving your cat alone when they are sleeping. If the cat is subjected too often to SCAT, they may need a beta blocker to keep the epinephrine at bay. My advice: Let sleeping cat's lie, their heart health depends upon it.

Hey, here is a poem I made up that fits this photo.

Oh the Christmas tree is pretty

All decorated in silver and green in the city

Attracting the family cat to climb on up it

Their claws accidently ripping open the presents that sit

Under the tree

So when your cat tells you

I hope all those presents are for me

You better make sure to see

At least several are for he.

Because never under estimate the narcissist cat he be.

I like this photo because the yellow and pink roses are beautiful and really bring out the cat's look of happiness. The photographer cloned in some sky behind the cat which I think accentuates the green hue in the cat's eyes. And what I think is outstanding is the perfect striping on the cat's face. This is definitely a healthy looking cat that is well taken care of. This cat is not lacking for love either. This photo could be used in the floral industry to promote cat love day and should be captioned "*Show your cat you love them with roses today!*" Unfortunately, since rose bushes have thorns, cats can't climb them, so giving roses is the next best thing. Meow!

Of course there has to be the token Christmas card with my photo featured. This was a big hit the year it was sent out by my owner.

Love it! That look says it all. I know the captioning is directed at humans, but I am the same way in the morning as I don't want to meow until I get my milk. I need to put this up on my cat bed before I hit the sack because the other cats are meowing too early in the morning for me.

This is an example of when Christmas cards with cats are all wrong. Cats don't drink eggnog, and do not eat candy canes. And they definitely are not friends with an anorexic mouse. And what's with the stoned look on the cat? Drinking a little too much of that liquored up egg nog fake cat?

This is a Halloween cat mask that has gone horribly wrong.
Looks like some retro psychedelic cat face from the 1960's. And what's with putting it on some kids face?
And you call those pumpkins? They look like blobs of molded canned cat food dyed orange for Halloween.
And that is no pumpkin patch. Look more like a catnip patch gone awry. Oh let me go roll around in that! Purrfect.

Is this above photo another example of psychedelic cat art?

I hope not! I can't look anymore!

The boy with the ADHD is grown up and letting the kittens sleep.

And see, he is not holding me in a neck squeeze.

And finally, we got to feature the token bird attack. I was outside and saw a robin sitting in a nest in a tree. The nest was out of reach for my paws so I climbed up the trunk. Well, mama robin heard me a coming and let out a yell for her male counterpart. She also came after me with her talons and beak. Now even though robins are small they got some power in those bodies. She and her buddy were pecking their beaks upon my fur in the tree. Since I do not take too kindly to being beaten on by some wild bird, I high cat tailed myself on out of there. And then I heard swooshing over my head as I was running. I looked up to see one pissed bird diving down at 50 miles per hour onto me. Praise the cat Lord my owner opened the front door and I escaped to the inside free of talon scraps!

"I hope you enjoyed *The Cat Chat Forum*. You know I am loved by my human masters. That is why my cat family is featured in [Squeaky's Sassy Cat Tales](), the next cat book by Laura. My cat brethren Squeaky, Bambi, and Rex, the family dog, and I tell our stories about our wonderful cat life living in Master Laura's household. Let her know you want to read it, she wants to hear from you so she can get it published".

Sincerely Tommy the cat. Cat peace and love to you all!

www.ingramcontent.com/pod-product-compliance
Lightning Source LLC
Chambersburg PA
CBHW041533040426
42446CB00002B/70